Josep

Published by
JOBECO BOOKS
BOX 3323
HUMBLE, TEXAS 77347-3323

Cover Photographs Courtesy "Collections of Greenfield Village and the Henry Ford Museum".

Printed in the United States of America
First Edition 1981
Library of Congress Catalog Card No.: 81-85407
JONES, JOSEPH C. JR.
ICE BOXES

ISBN: 0-9607572-0-1

D. Armstrong Co., Inc.
Printers & Publishers
Houston, Texas

TO BEVERLY, SHIRLEY, KEVIN,
RONALD, AND BRUCE

Contents

Preface

Probably the first question one would ask upon becoming aware of this book is — why would anyone write a book devoted to the subject of ice boxes? There isn't a particularly clear cut rational answer. The project simply evolved. I would, however, like to share with you its evolution so that you understand the source of my interest, motivation, and enthusiasm for the project.

I was brought up in a family that was somewhat interested in saving heirlooms — particularly furniture. My mother-in-law devoted enormous time and energy to finding and refinishing antique furniture. So, it was only natural that Bev and I were interested in older furniture when we set up housekeeping. It was a necessity too. We couldn't afford new furniture! We shopped for bargains — trying to obtain old furniture that was "priced right" and needed refinishing.

Over a period of several years we've collected odds and ends (antiques or collectibles) that make up the predominate part of our furniture.

Several years ago we were transferred to Houston, Texas. We spent several months getting acquainted with antique shops in the area that carried an inventory that was of interest to us. It was during this timeframe we became intrigued with ice boxes and agreed to purchase one — a porcelain lined box without manufacturer identity and needing minimal refinishing work.

Just a few weeks later we found a Baldwin (Burlington, Vermont) that was covered with several layers of weathered, cracked, and peeling paint. In spite of the horrible superficial appearance, it obviously had great lines, ornate carvings in the wood, and brass hardware (hinges and locks). We had to have it!

We reasoned that since it was so much better than the first acquisition, we'd refinish both of them and sell the first one. Well, things haven't worked out that way at all. We still have ice box number one plus several others.

Why this interest in ice boxes became so strong after moving to Houston is beyond me. Certainly we had been exposed to them while living in other parts of the country. Perhaps it's a subliminal reaction to the fact that Houston is a city without heritage, relative to cities we've lived near in the East — Philadelphia, Boston, and New York. Ice boxes do represent a bit of Americana and a link to the past.

The investment opportunity for ice boxes (and other collectibles) without question had some influence on me. I firmly believe in the right and, indeed, the obligation of individuals to save and invest during their productive years. Antiques and collectibles have and should continue to be excellent investments. Prices have continued to escalate over the last few decades with little fluctuation that occurs with other investments such as commodities, stocks, and precious metals.

After acquiring several boxes and learning several refinishing techniques peculiar to ice boxes, I developed a growing desire to write a book on where to find and how to restore them. However, in the process, I learned the fascinating history of the ice industry — its implications on the U.S. economy, importance to food preservation, etc. This ultimately lead to a book with much broader treatment of this subject.

Acknowledgements

Dozens of people — those associated with historical societies, universities, museums, and libraries, as well as several individuals, have provided background material and support in preparation of this book. Although it is impossible to list by name everyone who contributed, their help is greatly appreciated. I would in particular like to thank the following people and institutions: Clement G. Vitek and Nancy Oppel, The Sunpapers, Baltimore, Maryland; Jane Stevens, Chicago Historical Society, Chicago, Illinois; H.O. Williams, Flat River Historical Society, Greenville, Michigan; Kathy Vocelka, Houston, Texas; Donna Braden, Greenfield Village and Henry Ford Museum, Dearborn, Michigan; April Schwartz, Minnesota Historical Society, St. Paul, Minnesota; Frances Forman, The Cincinnati Historical Society, Cincinnati, Ohio; The Farmers' Museum, Inc., Cooperstown, New York; Nancy Kessler-Post, Museum of the City of New York, New York, New York; Druscilla Null, Museum and Library of Maryland History, Baltimore, Maryland; Alberta Rommelmeyer, The Chippewa Valley Museum, Eau Claire, Wisconsin; Celene Idema, Grand Rapids Public Library, Grand Rapids, Michigan; Helen Kahn, Indiana Historical Society Library, Indianapolis, Indiana; Peter Drummey, Massachusetts Historical Society, Boston, Massachusetts; Isabella Athey, Maryland Historical Society, Balitmore, Maryland; The University of Vermont, Burlington, Vermont; Carole A. Loll, Ella Sharp Museum, Jackson, Michigan; Gordon Davenport, Kingston, New York; and my parents, Mr. & Mrs. Joseph C. Jones.

HISTORY OF REFRIGERATION

History of Refrigeration

As we fill tumblers with ice cubes from the ice maker, sneak a dish of ice cream from the freezer compartment or simple enjoy a glass of cold milk, seldom do we give much thought to this marvelous invention — the refrigerator. It is but another convenience and a major technological break-through we simply take for granted. And, yet refrigerators, as we know them today, have existed for only a few generations during this century.

Preservation of food through crude refrigeration was an early science for which there are innumerable references. The word refrigeration is outdated in that it is derived from the Latin word frigus, meaning frost. Refrigeration can be defined as: cooling of a body by the transfer of a portion of its heat to another, which needs to be, of course, a cooler body. The flow of heat from warmer bodies to colder bodies is called heat transfer. The process causes the temperature to rise in the colder body and decreases in the warmer body as the heat is transferred. Refrigeration, as a means for food preservation has had a significant influence on the progress of civilization.

Early History

Alexander the Great, King of Macedon (B.C. 336-323) had trenches dug which were then filled with snow. This was used to cool hundreds of kegs of wine which were given to his soldiers on the eve of battle. His troops, of course, were able to do battle the following day without much care as to the consequences.

Nero, the Roman Emperor (54-68 A.D.) had his wines cooled by snow brought down from the mountains by slaves.

The Indians knew the fundamentals of refrigeration. The northern tribes had fish and portions of meat frozen during the winter.

Francis Bacon, a prominent English philosopher and statesman (1561-1626), realized the importance of refrigeration and its potential impact on mankind. To this end, he attempted many experiments. It is reported that one day in 1626, Bacon went to London and while driving near Highgate, became intrigued with a desire to learn whether snow could act as an antiseptic and preservative. He stopped the carriage, purchased a fowl, had it killed and cleaned, and assisted in stuffing it with snow. Unfortunately, no conclusions were developed from this experiment in that Bacon contracted a serious cold which resulted in his death a very short time later.

There is ample evidence that the use of ice for refrigeration, although seemingly insignificant, had an important influence on the history of the world. Because of the lack of knowledge and experience, ice originally was used as a luxury. Greater demand for sanitation, more effective food preservation, and transportation ultimately caused it to become a necessity.

American Ice Harvest

The demand for ice as a necessity for food preservation gained momentum. The more general use of ice created the "natural refrigeration" period during which natural ice harvesting from ponds, lakes, and rivers progressed from a small localized activity to a large industrialized business producing enormous quantities of square-cut block ice quickly and efficiently. Although experiments with ice making machines in the mid 1800's looked promising, they were at first ignored as being non-commercial. Large crops of natural ice were generally available.

Impact of Technology

Ultimately, technology won out and the natural ice trade faded into oblivion. The inventors who first were successful in making artificial ice by modern methods were Americans. In 1834 Jacob Perkins, of Newburyport, Massachusetts, who was at the time residing in England, obtained a British patent. The ice making machine patent was based on a principle of the evaporation of sulphuric ether under an air-pump. His ether-compression machine was the forerunner of compression-type equipment.

Professor A.C. Twinning of New Haven, Connecticut procured a British patent in 1850 and a United States patent in 1853. In 1855 he had a machine in operation in Cleveland, Ohio that produced 1,600 pounds of ice in a twenty-four hour run.

In 1851 Dr. John Gorrie of New Orleans patented a machine for producing ice by compressing and expanding atmospheric air.

In 1848 Ferdinand Carre of France contrived an original process for employing ammonia. In 1865 he patented an ice machine and in 1867 at the French International Exposition it produced six tons of ice daily.

Interest in ice making equipment gained momentum during the latter half of the 1800's. Uncertainity of supplies, great fluctuations in price, and sanitation standards made natural ice vulnerable to the "new" artificial product. Installation of ice making and refrigeration equipment progressed rapidly in many population centers of the United States. Ice making equipment became increasingly important for the long distance shipment of perishables such as fruits, meat, and dairy products. The production of "plant ice" gained such inroads on the natural ice harvest that by the 1920's little remained of the industry. Giant ice houses stood vacant along the Hudson, Kennebec, and other northern rivers. Symbols of an industry that in a matter of decades experienced both rapid growth and importance to the American economy and unprecedented obsolescence due to technology, ingenuity, and individual enterprise.

ICE HARVEST

Ice Harvest

Little was written regarding the natural ice crop prior to the 1800's. Ice was harvested primarily by farmers for their own use — especially for the storage of meat and dairy products. The early ice cutters had very simple crude equipment consisting only of axes and cross-cut hand saws. Harvesting was time-consuming and very inefficient. There were ice depots in some cities and towns supplied by farmers who cut ice from nearby ponds. However, transportation and storage facilities were woefully inadequate. The ice was generally sold to consumers as irregularly shaped fragments in bushel baskets.

Photograph Courtesy The Harry T. Peters Collection
Museum of the City of New York

Winter in the Country/Getting Ice
Painted by G.H. Durrie, lighograph by Carrier & Ives, 1864

In rural areas, where it was possible to harvest ice in the winter, ice houses were owned by individuals such as farmers or cooperatively by a few families living close enough together to conveniently use one in common. Farm ice houses were generally long low structures located near the house and other buildings.

Filling the house took place in the bitter cold winter. The ice was hauled by bobsled from a lake or river — assuming there was sufficient snow. There was a big advantage of using sleds. The wagon box was closer to the ground than a wagon with wheels. Ice cakes weighing 100 pounds or more could be loaded into the sled just a little easier.

Cutting, loading, transporting, and storaging the ice was slow, tedious work. Men dressed in horsehide coats, heavy mittens and boots cut the ice into cakes, drug them out of the water with tongs, and loaded them on a sled for transportation to the ice house for storage.

In the house, each layer of ice cakes were arranged in rows. A space of at least twelve inches was left around the outside edge and packed tight with either sawdust or straw.

Photograph Courtesy The Baltimore Sunpapers
This ice house on a dairy farm near Cockeysville, Maryland provided ice for dairy and kitchen use. Ice was harvested from a pond a mile and a half away. The building's rock-lined walls go 18 feet under the ground.

AN ICE HOUSE (?) AT HARPERS FERRY

Harpers Ferry, W.Va., now a National Historical Park, was first settled by Peter Stephans in 1733. He set up a primitive ferry service there. Fourteen years later Robert Harper, the man for whom the town is named, took over Stephans ferry operation. Sometime between then and John Brown's infamous raid in 1859, this building was constructed. National Park Serivce personnel believe it to be an ice house. They found two feet of sawdust on the floor and a built-in floor drain.

This ice house is located at the Franklin D. Roosevelt home (National Historic Site) in Hyde Park, New York. James Roosevelt, FDR's father, built a pond specifically for harvesting ice in 1881. Ice was harvested each winter from 1882 to 1941.

A thick blanket of straw or sawdust, usually at least twenty inches, was placed on top of the stack to insulate the ice from the summer heat. In spite of all the precaution to prevent melting, it was common for half of the ice to melt. It was, therefore, necessary to harvest nearly twice as much as required to compensate for the melting loss.

Commercial ice harvesting in the United States took place in the months of January, February, and March. As soon as the ice was thick enough, operations commenced. Thickness of the ice varied. For instance, in Southern New York on the Hudson River, in Pennsylvania, Maryland, and throughout the Ohio River Valley, ice would commonly be cut when it was 6 inches thick. This was due to the uncertainty of the weather. Further north it was common to wait until the ice was ten to twelve inches thick. In Maine, much of the ice was 15 to 30 inches thick.

To begin the harvest, the ice was scraped when covered with snow. If the surface was rough it would also be planed.

Gifford-Wood Co. Economy Snow Scraper.
(Patent applied for.)

From Gifford-Wood Co. Catalog
Gifford-Wood Co. economy snow scrapper as advertised in their September 1923 catalog.

After clearing the snow, the area was "prospected" to determine the best place to begin cutting. Holes were drilled and a measuring rod inserted to determine the thickness. Preference was given to that part of the field where the ice was thickest and upstream from the ice house, if on a river, in order to make best use of the stream current in floating detached ice to the house.

From Gifford-Wood Co. Catalog

A field planer with cutter bar and flanged teeth.

Once selected, the field was immediately marked off into squares. This was done with a horse-drawn marker steered by plow handles along a straight line marked on the ice. It normally cut to the depth of 2 inches. At the end of the field the marker was turned around, a sliding guide placed in the line just made, and then drawn back across the field cutting a new groove. As soon as the field was lined off in one direction, a new set of lines were run at right angles which divided the field into perfect squares.

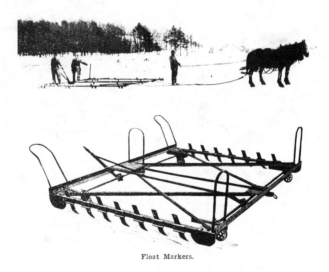

Float Markers.

From Gifford-Wood Co. Catalog

A float marker constructed of a steel frame and seven cutting teeth on each side and drawn by two horses. After making the first line with a hand plow, one side of the float marker always cut in the groove previously made by the other side. Since each side cut one inch, the resulting grooves left by the marker were two inches deep.

The ice cutter or ice-plow was the next implement used in the harvest. The plow was designed to cut to about two-thirds the depth of the ice so that the blocks could be detached with hand tools. The horse-drawn plow would run through grooves cut by the marker.

Graphics From H. Pray Brochure: Clove, New York
The brochure states "It will cut from 20 to 40 tons an hour, cutting the ice blocks of uniform size, which saves much labor when packing. The plow is about 4 ft. 4 inches in length, teeth and runners are adjustable and are made of the best steel".

Once blocks were detachable, a channel would be opened from the field to the ice house. In some cases, the channel would be open to loading platforms adjacent to trains for immediate shipment to ice houses closer to the market. The procedure for harvesting the field was to detach a "sheet" of about 12 blocks of ice and send them down the channel. The sheets were either pulled by a team of horses or poled along by an ice cutter using a tool called a "hook".

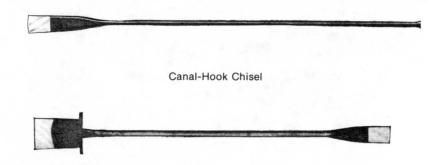

Canal-Hook Chisel

Breaking Bar

Photograph Courtesy Gordon Davenport

Binnewater Lake Ice Co.
Ulster County, New York
Large sections of ice have been scored and are being cut through, broken off, and floated into the channel of clear water.

Binnewater Lake Ice Co.
Ulster County, New York
The ice blocks were transported by an endless-chain elevator driven by 25-35 horse-power engines to either an ice house or freight train.

Photograph Courtesy Gordon Davenport

Binnewater Lake Ice Co.
Ulster County, New York
Specially designed freight cars being loaded with ice for delivery to large warehouses in nearby New York cities.

When the sheet arrived at the ice house, it would be separated by light-breaking bars (see illustration) into single blocks for elevating into the ice house. The ice was packed away in the house in layers with slight cracks between the blocks. The cracks allowed the melting water to run off and at the same time prevented the cakes from freezing into one massive ice block.

Interior of house — note the cracks between the ice blocks.

After the ice house was filled, the top was covered with straw, hay, or sawdust. This was done to insulate the ice from warm spring and summer temperatures. Even with best insulation effort, there was melting loss that ranged from 10 to 25 percent.

Harvesting ice was very labor intensive. For instance, to fill a large ice house with a capacity of 25,000 tons would have required minimally 100 men and 10-12 horse teams. The harvest was usually completed in 15-30 days.

Most of the ice houses were built of wood, sometimes of brick, and commonly located at the edge of a lake or river where transportation was readily accessible to move ice to markets.

19

THE NATURAL ICE TRADE

The Natural Ice Trade

An active commercial ice business did not exist until an enterprising merchant began cutting blocks of ice from a pond in New York City and shipping it to Charleston, South Carolina in 1799.

Nathaniel Jarvis Wyeth from Boston is credited for greatly increasing the scale and efficiency of ice harvesting. He developed interest and expertise while managing a family-owned hotel (Fresh Pond Hotel). The hotel had an ice house that was filled each winter for use during warmer months. During the winter he worked on improved ice harvesting methods while cutting ice for storage. In 1825 he devised an ice cutter — also known as an ice plow.

Feeling the need for a career change, Wyeth agreed to exclusively supply Fredrick Tudor, the "Ice King", with ice for his growing markets. The relationship lasted for many years. During this time, he invented and patented other tools that advanced large scale ice harvesting.

In the early 1800's, an export trade developed in natural ice. Frederic Tudor (1783-1864) became known as the "Ice King" from his success in developing the business of shipping ice from Boston to the West Indies. He was the son of a prominent Boston family. He did not, like his brothers, attend Harvard College. Instead, he went into business at the age of thirteen. At the age of twenty, he, with the help and support of a brother and cousin, sent a cargo of ice to Martinique. The vessel arrived at Saint-Pierre in March 1806.

For the next fifteen years, he persisted in developing the business in spite of ridicule by friends, being in debt, and sometimes in jail. By 1821, he had established a business in Havanna, Charleston, and undertaken a venture in New Orleans.

Tudor, through much experimentation learned how to ship ice with the least possible loss, utilizing speedy clipper ships. He devised a structure that would "keep" ice in warm climates and ultimately succeeded in making the use of ice a necessity in these cities. Although confronted with much competition, he succeeded to become rich because of his ruthless methods, fanatical belief in the business, and determination. He also succeeded in establishing business in the Far East. In May 1833, the first cargo was sent to Calcutta. This made possible a worldwide expansion of his business.

This trade was invaluable to the city of Boston. The number of tons of ice shipped from Boston increased from 130 in 1806 to 146,000 in 1856. In 1856, 363 cargos were sent to fifty-three different places in the United States, the West Indies, The East Indies, China, the Philippines, and Australia.

A commercial ice business developed rapidly in other cities too. In both Baltimore and Washington, ice from near and far was stored in ice houses on the waterfront and later distributed by horse-drawn wagons.

Photograph Courtesy The Sun Papers, Baltimore, Maryland
A wagon of an independent dealer, Lawrence McDonald, delivering ice to row houses in Southeast Baltimore.

Large quantities of ice were cut and stored along the Susquehanna River, north of Baltimore. It was towed to the city by barge the following summer. The men who cut and harvested the ice in the winter also drove the ice wagons during the summer.

In the summer, ice was delivered six days a week. A second round was made late on Saturdays to tide the housewife over Sunday. There were no Sunday deliveries. Wagons were not allowed on the streets that day. In the winter, many customers discontinued purchasing ice. Instead, they used a wooden box nailed to the outside windowsill to store milk, butter, and other perishables. Even during the winter in Baltimore a limited market for ice existed — particularly with meat stores, restaurants, and saloons. The saloons, especially, used ice year round to assure customers of a cold drink.

Photograph Courtesy The Sun Papers, Baltimore, Maryland
Ice wagons were familiar sights until the early 1920's. This one is believed to have been photographed in Baltimore in 1910.

Photograph Courtesy The Sun Papers, Baltimore, Maryland
Iceman George Younger holds a box of crushed ice and his assistant, Nick Keller, holds a block of ice while making deliveries to saloons along a section of Eastern Avenue in Baltimore known as The Bowery. The picture was taken around 1912.

The demand for ice necessitated bringing in supplies from northern states to supplement local production.

The first ice brought to Baltimore from Maine arrived in 1820. During the 1800's a large fleet of schooners were used to carry coal from Baltimore and Hampton Roads to northern ports. The return cargos would frequently be ice from Maine to Baltimore and Washington. The ice would be harvested during the winter on the Kennebec and other freshwater Maine rivers and stored in ice houses on the riverbanks.

Photograph Courtesy The Sun Papers, Baltimore, Maryland
The four-masted schooner Gen. E.S. Greeley unloading ice at an ice house on Covington Street in Baltimore around 1900. Schooners commonly carried coal from Baltimore and Hampton Roads to northern ports and ice from Maine to Baltimore. Maine ice was considered to be of superior quality.

Maine was especially well suited for commercial ice harvesting. Her rivers were famous for the purity of their water, the long cold winters insured ice that was thicker and denser, and pine sawdust, an unwanted by-product of the Maine sawmills, was used for insulation.

By the late 1800's, Maine had become the leading ice harvesting state with 90% of its production going to other states and countries. Cities such as New York, Philadelphia, and Baltimore were particularly dependent on Maine ice. An enormously successful export business occurred in the 1879-1880 season. It is described in *The Ice Industry of the United States,* by Henry Hall, as follows:

"The harvest had been proceeding quietly through January and February, when the word came from New York that the Hudson ice crop had failed, and that wholesale prices had risen in New York to $4 and $5 a ton. Ice was from 15 to 20 inches thick on the Kennebec, free from snow ice, smooth enough to cut without planing, firm, hard, and brilliant. It could be cut for 20 cents a ton; loaded on vessels for 50 cents; freighted to New York for 50 cents; and landed there at a cost of about $1.50 per ton. The advices received were so favorable that all the icemen of the Kennebec were thrown into a state of great excitement. The companies all prepared to cut immediately as much as they could handle; and a number of new concerns promptly invested a good deal of money in tools and went into business. Every idle workman along the river was employed and put to work. Shipyards and sawmills were applied to for sawdust to pack the ice; the demand was so large and the supply so inadequate that the sites of old sawmills were hunted up in order to dig out sawdust several years old.

The portions of the river where cutting was done were covered with an army of 4,000 men and 350 horses, and work was presented day and night.

For fifteen years there had been no such business excitement in Maine."

The ice business was also of substantial size in New York State. During this period, large quantities of ice were cut and stored on the Hudson River for use in New York City the following spring and summer. Ice dealers were also served with ice cut by farmers from lakes in the lower counties of the state such as Rockland Lake near Nyack.

In 1880 there were approximately 160 large commercial ice houses located along the Hudson River in communities such as Newburgh, Poughkeepsie, Peekskill, Rondout, Hudson, and Albany.

However, this storage capacity did not meet the full demand of New York City. It was estimated that the city and surrounding

The Ice Cart, c.1840, by Calyo

Cutting Ice on the Hudson River

communities consumed 1,500,000 tons of ice per year. Although 2,000,000 to 2,750,000 tons were harvested during a good winter, over 50% was lost from meltage and waste in transit. The deficit was supplied by the Lake Champlain region and Maine.

The natural ice business developed in other regions of the country as well. Beginning in Chicago about 1840, Alson Smith Sherman operated the first commercial ice house. The ice house, located along the Chicago River, was acquired by Sherman as payment for construction work from a customer who was short of cash. The ice business flourished.

Alson S. Sherman

Alson S. Sherman, the first Chicago ice dealer.

Chicago was a booming city growing from 30,000 in 1850 to 109,000 in 1860. With the growth came pollution of the Chicago River and regulation by the Common Council and City Physician. Sale of ice harvested from the river was prohibited. The result — ice was harvested and imported from northern Indiana and the Fox River Valley. The ice was shipped in via railroads.

The Willow Spring Ice Houses.

Demand for ice skyrocketed, not only as a result of population growth, but also the growth of two primary industries — breweries and meat packers. Breweries needed to chill beer during the fermentation process.

Brewing, which had previously been conducted only in the cool season, could now, with the availability of ice, be carried on throughout the year. This industry consumed enormous quantities of ice for manufacturing and storage.

Meat packers had to refrigerate fresh meat in boxcars while enroute to consumers hundreds of miles away. Armour and Swift and Company began to develop their own ice harvesting and storage facilities in the late 1800's.

By the end of the 1800's, the Chicago Board of Health had dictated such tough standards that ice cut from waters of Indiana and Illinois were considered polluted. The industry shifted to Wisconsin. The ice was transported to Chicago by rail and lake schooner. It is reported that in 1880, 1.1 million tons were shipped to Chicago. Although supplies from Wisconsin were less vulnerable to mild winters, which was bound to mean shortages the following summer, they were much more expensive. Consumers complained. Merchants and saloon keepers protested. Newspaper editorials proclaimed that excessive profits were being made on a basic commodity.

SULLIVAN ICE COMPANY,

Private Families and Others supplied with the best Table

RESIDENCE:
COR. BOSWORTH AVE. AND ROSCOE ST.

224 Lincoln Ave.

OFFICE:

Families a Specialty.

ICE HOUSES AT SILVER LAKE, WIS.

TELEPHONE 3408.

Graphic Courtesy Chicago Historical Society. ICHi-16100

Sullivan Ice Company

ICE !

The Undersigned will commence delivering ICE, Tuesday, May 26, '68, at the same prices charged by the CHICAGO ICE COMPANIES, viz:

105 lbs per Week,	- -	$ .80
140 " " "	- -	1,00
210 " " · "	- -	1,50

The ICE will be delivered inside of Front Gates. When carried through Yards and put in Ice Boxes or when delivered away from the ROUTE an additional CHARGE WILL BE MADE.

BILLS will be presented for payment the first of every month.

J. W. MERRILL.

Graphic Courtesy Chicago Historical Society. ICHi-06222
A notice by J.W. Merrill of his intention to deliver ice at competitive prices (May 1868).

The Knickerbocker Ice Company by 1898 had bought up most of the ice companies in the Chicago area. Although very powerful, it was also very vulnerable. It's monopolistic operation was criticized by the press and politicians. In 1901, the Illinois Pure Food Commission condemned Chicago ice for being "cut from impure ponds with sewage, decayed animal and vegetable matter". New competition surfaced. Artificial ice, made by Consumers Ice Company, with purified water, became economically competitive. Consumers Ice Company not only made inroads on Knickerbocker Ice Company's natural ice business, it ultimately purchased Knickerbocker.

Photograph Courtesy Chicago Historical Society
Photographer: Fred Tuckerman, ICHi-05141
An iceman delivering his product in Chicago (1908).

In the southern states where ice did not form naturally, it was imported from the north. A significant part of Frederic Tudor's ice trade was with the southern states. The principal consuming cities were Savannah and New Orleans. Other cities of significance were Wilmington, North Carolina; Charleston, South Carolina; Jacksonville and Pensacola, Florida; Mobile, Alabama; and Galveston, Texas.

The cost of transportation plus loss through melting and waste made ice very expensive, ranging from $20 to $75 per ton. Only the rich could afford this luxury. The average person in the southern states simply did without refrigeration as a form of food preservation.

ICE BOX MANUFACTURERS

Ice Box Manufacturers

Although ice had been used for centuries to preserve food and cool beverages, it was not commonly used in households until the 1800's. By the end of the century it is estimated that approximately one-half of the annual natural ice crop was used for home food storage. That evolution was the result of many developments including: a large scale ice harvesting industry to help insure supplies at economical prices, a distribution system that provided for home delivery in most large towns and cities, and the availability of ice boxes (refrigerators) that incorporated features based on a better understanding of refrigeration principles (i.e., insulation, the need for air circulation, etc.).

Ice boxes were used in American homes as early as 1830. It was initially believed that saving the ice was paramount. Therefore, it was quite common to wrap the ice in a blanket. The result — the ice was saved and the food spoiled.

Most ice boxes were lined inside with zinc, slate, porcelain, galvanized metal, or wood with a wall of insulation of either char coal, cork, flax straw fibre, or mineral wool. The outer casing was generally made of oak. Pine and ash wood were also utilized.

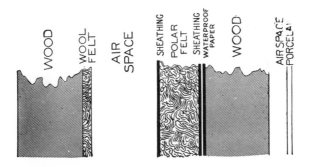

A sectional view of one wall of a Leonard Cleanable Porcelain Lined Refrigerator.

Oak was a very popular wood for making all kinds of furniture, including ice boxes, at the turn of the century. There were huge oak forests, particularly in the midwest, available for harvesting. Oak lumber could be kiln dried and used almost immediately. Additionally, oak was considered to have a tough grain, to be strong, and good for rigid construction.

By the 1850's the importance of air circulation for efficient cooling was realized. A system was patented in 1856 based on ice being placed at the top of the box and air circulating around it.

The principle of refrigeration by ice is comparatively simple. The air inside the box, flows to the ice, causing it to melt and the air to become cooler. This cooler air drops to the bottom of the refrigerator and displaces warmer air. As the warmer air rises it takes heat from the food in the lower levels. Circulation continues uninterrupted unless the door is opened.

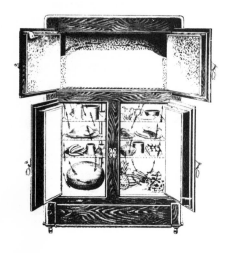
Circulation of cold air in an overhead ice system.

Circulation of cold air in a side ice system.

There was an enormous range in ice box sizes. The ice holding capacity varied accordingly, ranging from 50 to 125 pounds. Virtually everyday during warm weather, the ice man made his rounds to service customers. The ice man, along with his horse and wagon, became part of the summer street scene in every town and city. In spite of the necessity of his goods and services he was not always welcomed in the house as noted in the following advertisement.

McCray advertisement — July 1898

A block of ice would last a day or longer — if the doors were not opened too frequently. Water from the melting ice drained into a pan that had to be emptied often. Another alternative was for it to empty into a pipe that lead to the outside through the wall.

The Drainage System

A Bohn Syphon Refrigerator
Manufactured by White Enamel
Refrigerator Co.

The drainage system of most refrigerators continually got clogged. The drain pipe and trap of this system was located in the front where it was easy to get at. The manufacturer stated: "All it needs is the occasional removal of the drain pipe, run a stream of water through it, clean the discharge pipe in the bottom of the refrigerator with a cloth, put back the drain pipe, and the whole thing is perfectly sweet and clean, with absolutely no chance for foul matter to accumulate, and the whole operation has taken but a few seconds".

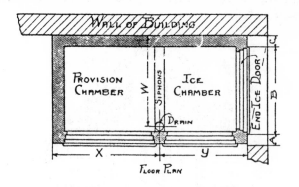

WALL OF BUILDING

PROVISION CHAMBER	ICE CHAMBER	END ICE DOOR

W — SIPHONS

DRAIN

C
B
A

X — Y

FLOOR PLAN

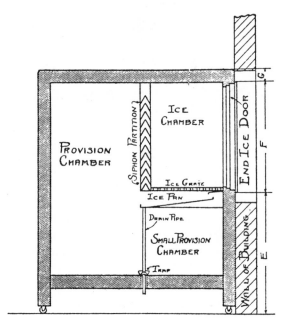

PROVISION CHAMBER

SIPHON PARTITION

ICE CHAMBER

END ICE DOOR

ICE GRATE
ICE PAN

DRAIN PIPE

SMALL PROVISION CHAMBER

WALL OF BUILDING

TRAP

G
F
E

SECTION

From 1916 Seeger Refrigerator Catalog

The company advertised "Keep the ice man outside and prevent the tracking of mud on your clean floors by having a Seeger Refrigerator with either an end icing door or a rear icing door". The above illustrations show the typical specifications for outside icing doors.

Early each day a cardboard sign was placed in the window to indicate the amount of ice to be delivered.

The sign indicated whether to deliver 25, 50, 75, or 100 pounds of ice.

Dozens of companies competed for the ice box market. Each of whom offered a variety of models. Following is a partial list of manufacturers:

Baldwin Manufacturing Co., Burlington, Vermont
Belding-Hall Refrigerators, Belding, Michigan
The Brunswick-Balke-Collender Co., Chicago, Illinois
D. Eddy & Son, Boston, Massachusetts
Elkins Refrigerator and Fixture Co., Elkins, West Virginia
Eureka Refrigerator Co., Indianapolis, Indiana
Grand Rapids Refrigerator Co., Grand Rapids, Michigan
Gurney Refrigerator Co., Fond Du Lac, Wisconsin
Hedand Hedenberg, New York, New York
The John C. Jewett Mfg. Co., Buffalo, New York
The Keyser Manufacturing Company, Chattanooga, Tennessee
L.H. Mace & Co., New York, New York
McCray Refrigerator Co., Kendallville, Indiana
McMichael & Co., Boston, Massachusetts
Metal Stamping Co., Jackson, Michigan
Minnesota Refrigerator Co., St. Paul, Minnesota
Monroe Refrigerator Co., Lockland, Ohio
Northern Refrigerator Co., Grand Rapids, Michigan
Ranney Refrigerators, Greenville, Michigan
Seeger Refrigerator Co., St. Paul, Minnesota

Simmons Refrigerator Co., St. Paul, Minnesota
E.H. Stafford & Bros., Chicago, Illinois
Peter A. Vogt, Buffalo, New York
Jos. W. Wayne, Manuf'r., Cincinnati, Ohio
White Enamel Refrigerator Co., St. Paul, Minnesota
Wilke Manufacturing Co., Anderson, Indiana
The Wisconsin Refrigerator Co., Eau Claire, Wisconsin

The latter part of this chapter is devoted to brief histories, advertisements, and product literature of several companies that manufactured ice boxes. The companies truly represent American entrepreneurship — individual initiative and ingenuity coupled with the inherent risks of a free marketplace. During their heyday, the owners of most companies prospered. Thousands of workers were employed in this industry throughout the United States. But technology (the electric home refrigerator) caused the ice box to become obsolescent very quickly in the early 1900's.

In 1913, the "Domestic Electric Refrigerator" was marketed in Chicago. In 1914 Kelvinator introduced an electric refrigerator. Interestingly, it was recommended that the motor and compressor be housed in the cellar and linked to the ice box in the kitchen by a pipe. This was because the refrigerant, sulphur dioxide, was quite odoriferous and best kept as far as possible from the kitchen area.

It was not only the improved technology that caused the demise of ice box manufacturers. The companies making the new electric refrigerators were aggressive, innovative marketers. Canned sales pitches were developed for Frigidaire salesmen. Following is one from a 1923 "demonstration" book for calling on the prospect at home:

Approach and Album Demonstration
At the Home

"How do you do, Mrs. Prospect? _____ is my name and I represent the_____ _____which sells *Frigidaire* in this city.

42

(Salesman should immediately try to enter the house, by acting as if he expected to go in, taking a step forward.)

"Mrs. Prospect, I sent you some booklets describing *Frigidaire,* the *Electric Refrigerator.*

"I came out this morning to tell you more about electrical refrigeration and what Frigidaire can do for you and your family.

(If salesman has not gained admittance, he makes another start and smilingly says, "Mrs. Prospect, I want to show you these pictures which very clearly tell the story of *Frigidaire.*"

Salesman proceeds to give *Album Demonstration.* In some instances it may be the best policy to go at once to refrigerator, giving album demonstration later.)

"Mrs. Prospect, will you let me see your refrigerator?

(Salesman follows prospect to the kitchen. While in the kitchen he takes a thermometer out of his pocket and puts it on top of the refrigerator to get the room temperature.)

"About how much ice do you use each week, Mrs. Prospect?

"Are you entirely satisfied with this size refrigerator?

(Salesman takes out rule and measures outside of refrigerator.)

"About how long have you had this refrigerator?

(Salesman makes note of general condition of ice box. Before proceeding any further, the Salesman reads the thermometer and calls the prospect's attention to the room temperature.)

"This is a tested thermometer, Mrs. Prospect, and you will notice that the temperature of this room is _____ degrees Fahrenheit.

(Salesman then places thermometer in the food compartment of the refrigerator.)

"May I ask how many people you have in your family?

(If prospect is a young woman, ask, "Are any of them small children?"

Salesman takes a step away from refrigerator and regards it for a moment.)

"Mrs. Prospect, as I showed you a few minutes ago, we find that the average ice box maintains a temperature

43

of about 55 degrees, and I think you will agree with me that this will keep food properly for only a short time.

(Salesman takes thermometer out of ice box and shows temperature reading to prospect.)

"The temperature in your refrigerator is_____degrees. This is slightly warmer than I expected. If you had a *Frigidaire,* the temperature would certainly be——degrees colder than you now have in your ice box. You would be able to keep your food properly, and have the conveniences just as I pictured to you a few minutes ago.

(If salesman decides that a complete *Frigidaire* is the equipment he should sell, he can proceed as follows: "I think I have obtained about all the information I need, and I am going to leave these illustrated booklets with you."

If salesman decides a B-R should be sold, he should proceed as follows: "Mrs. Prospect, in your case it would not be necessary for you to buy a new refrigerator. The one you have will be satisfactory for mechanical refrigeration and we can install *Frigidaire* in it. With *Frigidaire* installed in your present refrigerator, you would obtain all the advantages that I have shown you and it will be satisfactory in every way.")

"Won't you please talk this matter over with your husband tonight as, in all probability, I or one of our men will call upon him tomorrow afternoon and tell him the benefits of owning a *Frigidaire.*

(Ask for husband's business address if it seems advisable.)

"I certainly hope that within a short time you will have a *Frigidaire* in your home. Good-bye Mrs. Prospect, and thank you for giving me your time."

(If prospect asks the price, give a quotation which will help you make the ultimate sale.)

The literature distributed by Frigidaire was also very effective in pointing out the inconvenience of ice boxes.

The new much-improved electric refrigerator, aggressive marketing by companies like Frigidaire, and the depression of 1929

44

The iceman does his best to serve us. Often the iceman does not arrive when we expect him. He sometimes tracks up the kitchen floor and causes annoyance. The delivery of ice in the ordinary way can never be looked upon as a very sanitary proceeding. The reason for all this has been the conditions, rather than the desires of the iceman.

Waiting for ice. This is a further annoyance that is often caused by the fact that some member of the household wants to leave, but cannot because the ice has not come.

No one has to stay home when you own a Frigidaire. While you are away, Frigidaire works automatically and makes ice which is always ready for use. It is very convenient to be able to leave at the week-end or for a vacation, with no worry about food spoiling, ice delivery, or other matters pertaining to the refrigerator.

From Frigidaire Booklet

all contributed to the rapid demise of ice box manufacturers. None of the companies make ice boxes and few even exist today. They either ceased doing business or were merged into larger companies.

The information needed to provide individual company histories ranged from being reasonably complete to virtually non-existent. It is my hope that someday another edition of this book will be published containing more complete information than I have been able to obtain thus far. And now, to information about the individual companies.

The Baldwin Refrigerator Co., Burlington, Vermont

Judson A. Baldwin of Shelburne, Vermont invented and made the first Baldwin dry air refrigerator in 1880. The following year the firm of Baldwin & White was formed. About 100 refrigerators were produced in 1881 by six to ten men working in the upper story of White's cheese factory. In August 1882 the Baldwin Manufacturing Company was organized and assumed the business and patents of the Baldwin & White partnership. Their factory was relocated near the Burlington railroad station for quicker access to the markets and their work force was increased

From McClure's Magazine, 1905

to forty men. By 1886, seventy-five to one hundred men were employed in several factories.

The company exhibited at leading fairs and expositions around the country where they took gold medals and other prizes. Their obsession with producing a superior product resulted in others attempting to imitate the Baldwin. The company obtained the sole right and title to use the word "baldwin" as applied to the refrigerator, and were granted injunctions against manufacturers and dealers using the term.

In spite of the initial growth and gust for excellence, the company was liquidated in 1936.

Belding-Hall Manufacturing Co., Belding, Michigan

Belding-Hall Manufacturing Company was formed in 1895 by a merger of two companies — Hall Brothers Manufacturing and Belding Manufacturing Company. The former company was organized in 1890 by Brinton F. Hall and three of his brothers.

"Collections of Greenfield Village and The Henry Ford Museum"
Belding-Hall refrigerator

They manufactured stove boards, sewing and card tables, and side-boards. The latter company, Belding, was established in 1875 as the Hembrook Manufacturing Company. It was reorganized in 1884 as the Belding Manufacturing Company. The company manufactured refrigerators exclusively.

By early 1920's Belding-Hall was reputed to have one of the largest refrigerator factories in the United States. It boasted a payroll of $130,000 per year and produced 5,500 refrigerators per month.

This company, like so many others, in the 1930's was severely affected by the depression. Its facilities were taken over by the Gibson Refrigerator Company and the Detroit Gasket Company.

Ladies Home Journal, May, 1910

The Brunswick-Balke-Collender Company, Chicago, Illinois

D. Eddy & Son, Boston, Massachusetts

Darius Eddy, a carpenter, began making refrigerators in the 1840's. His establishment became the largest one of its kind in Massachusetts. His famous "Upright Refrigerators" were shipped to many parts of the United States.

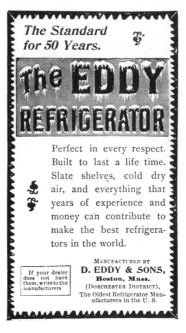

From McClure's Magazine, 1896

EDDY'S REFRIGERATOR.

The Best Refrigerator in the Market.

They are warranted to be packed with Charcoal, consuming LESS ICE than those that are not packed.

They are also provided with

SLATE STONE SHELVES,

And Ventilated in the most perfect manner. For sale at all the principal Furniture and House-Furnishing Warehouses. Also Manufacturers of **Improved Library Steps,** the most useful article one can have in the house.

D. EDDY & SON,

Harrison Square, Dorchester.

Store, 16 Bromfield Street, BOSTON.

From Boston Directory, 1869

50

Elkins Refrigerator and Fixture Co., Elkins, West Virginia

The first Elkins Cooler was built in 1905. Elkins produced refrigerators, counters, coolers, booths, and fixtures for commercial establishments such as grocery stores, restaurants, florist shops, etc.

Style No. 150

Ceiling Paneled Hardwood Front

THE front and finished side of this cooler is of hard wood, grooved and paneled as illustrated. Ornamented by the beveled plate mirror and the decorative steer head shield.

Carried in stock in varied sizes as listed below. As with style 450 this style is built in low heights with overhead ice chamber for rooms with a 9-foot ceiling. An exclusive Elkins feature. Other sizes to order as well as all sizes to order in C construction. See page 46.

Style No.	Outside Dimensions Front	Depth	Height	Approximate Ice Capacity	Approximate Weight
150	6'	6'	8'-10½"	2000 lbs.	3400 lbs.
	8'	6'	8'-10½"	3000 "	4200 "
	8'	8'	8'-10½"	3900 "	5000 "

From The Elkins Catalog, 1923

51

Elkins Florist Refrigerators

FLORISTS find in Elkins refrigerators the same qualities of efficiency, service and durability that have made them prominent in other businesses. They are designed for a special service and handle it capably, requiring a minimum of ice and attention.

These refrigerators are built to order only in the styles as illustrated. The cabinet work is excellent and the exterior wood is chosen in accordance with the finish desired.

Style No. 576F

This style illustrated above is finished in Elkins Standard Golden Oak. The interior is white enameled. There are three lights of glass in all doors and the ends, the outer glass being beveled plate. The ice door is faced with a beveled plate mirror. All compartments are fitted with heavily tinned wire shelves.

Style No. 576F — Front 88″, Depth 32″, Height 79″, Weight, approx. 1250 lbs. Ice capacity, approx. 600 lbs. Code — Floral.

	Size of Compartments			Door Openings	
	Height	Width	Depth	Height	Width
2 Display	47″	24″	24″	41½″	20½″
2 Display	21½″	24″	24″	18½″	20½″
1 Display	22″	27″	24″	18½″	20¾″
1 Ice	41″	23″	24″	38½″	20¾″

From The Elkins Catalog, 1923

Eureka Refrigerator Company, Indianapolis, Indiana

From McClure's Magazine, 1902

The Grand Rapids Refrigerator Company (Leonard Refrigerator Company), Grand Rapids, Michigan

The Leonard Refrigerator Company was incorporated in 1904. Leonard refrigerators, however, were sold long before an "official" company existed.

It was shortcomings of the domestic refrigerator (ice box) in his home and those sold in his father's store that ultimately led Charles H. Leonard into manufacturing refrigerators. The older refrigerators were "top icers" with immovable flues. If anything spilled into the flues, it created a tedious difficult cleaning job. One day in 1880, the maid in the Leonard home decided to quickly cool down a pail of hot lard by setting it on top of the ice. The ice melted unevenly causing the pail to tip over and spill lard down the flue. Mr. Leonard volunteered to clean the refrigerator. He found it was impossible to clean all of the lard out of the narrow space between the lining and the refrigerator wall through which the air circulated. The unpleasant task was the beginning of the Leonard Cleanable Refrigerator.

He was quoted as saying "That's no way to make a refrigerator — there should be some way to take out this inside lining".

The first patent was granted to C.H. Leonard in 1882 on his idea of building refrigerators with removable sides in the ice chamber. Charles and his brother Frank then decided to go into the business of making refrigerators. It was done on a part-time basis since they were principals in their father's household goods store — H. Leonard and Sons. The store hours, 7:00 a.m. to 11:00 p.m., did not leave much time to moonlight building refrigerators. They sub-contracted various jobs such as making the wooden cabinets and linings. The assembling was done in their own store building.

Modest success the first year encouraged the Leonard brothers to take an office and rent the second floor of a small building. Unfortunately, a fire wiped them out in 1883. Determination prevailed and the business opened at another location.

Getting the refrigerator company on its feet required a lot of hard work. It also took a lot of money. Charles Leonard must have been a master financier and borrower. His son, H.C. Leonard, who joined the company in 1897, said "It was all borrowed capital in those days. You know the refrigerator business is seasonal, you only sell them for a few months in summer, but you have to keep your plant going all the time. The only time we had any money coming in would be from the first of May, when folks began to want iced drinks, until along in the fall. Then it was up to father to borrow the money to carry on. He'd have some loans from every bank in town and most of them in Detroit, and then the next spring we'd pay up and start over".

THE HOME OF THE LEONARD CLEANABLE
THE LARGEST REFRIGERATOR FACTORY IN THE WORLD

THE plant of the Grand Rapids Refrigerator Company covers thirteen acres of ground. There are five railroad sidings to facilitate the handling of materials and finished stock. It includes Porcelain Enamel Works, Brass Foundry, Nickel Plating Works, Wire Works and Tinning Plant, in addition to the ordinary departments of a Refrigerator Factory. It has a capacity of 500 finished refrigerators per day. Be sure you place your orders with a factory that has the capacity to fill them.

From The 1914 Grand Rapids Refrigerator Catalog

From a very humble beginning of producing 10 to 15 refrigerators a day in the 1880's the company grew to a production of 2,500 units daily in April 1929. In 1933, company employees totaled 1,568.

In the interim the company was merged into a new firm called Electric Refrigeration Corporation and controlled by Kelvinator Corporation. Ice box manufacturing was ultimately discontinued.

Other mergers followed. Kelvinator merged with Nash Motors in 1937 to form Nash-Kelvinator. Nash-Kelvinator merged with the Hudson Motor Car Company in 1954 to become American Motors Corporation.

"Collections of Greenfield Village and The Henry Ford Museum"
Leonard refrigerator.

How he will talk! Great Gods! how he will talk! That is, your husband in praise of the sweet food that comes out of your *Leonard Cleanable Refrigerator.* Best in the world.

Applying the Government Test.

Place a new Green back in the door as shown and pull it out without tearing it, if you can.

The bill is yours if you succeed. This shows the doors are air tight.

No other Refrigerator will stand the test.

Be fair and foremost in the race And having won it hold your place... ...as the... *Leonard Cleanable Refrigerator has done.*

Illustrations regarding Leonard Refrigerators from a Grand Rapids Refrigerator Co. catalog.

Illustrations regarding Leonard Refrigerators from a Grand Rapids Refrigerator catalog.

Gurney Refrigerator Company, Fond DuLac, Wisconsin

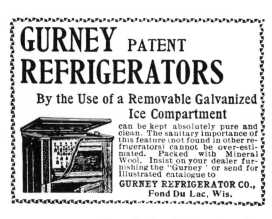

The Keyser Manufacturing Company, Chattanooga, Tennessee

From Ladies Home Journal, May, 1905

L.H. Mace & Co., New York, New York

The New York City directories include the following listings for this company:

L.H. Mace & Co. (Levi Mace), refrigerators,
- 1855-56 — 448 Houston St.
- 1860-61 — 20 Commerce & 111 Houston
- 1885 — 115 East Houston
- 1900 — 111, 113, 115 & 117 E. Houston & E. 150th N. River Ave.

No other information was obtained regarding this company.

McCray Refrigerator Company, Kendallville, Indiana

Hiram McCray and his sons Elmer and Homer were in the produce business. Their challenge was to cut down on the tremendous loss from food spoilage and, of course, to protect the health of the public too. Hence, their interest in refrigeration. In the years prior to 1890, they carried on their produce business and worked on experiments which led to the first patents for the McCray system of refrigeration.

The McCray Company was organized in 1890 by Elmer E. McCray based on a patent that had been granted to his father, Hiram McCray, in 1882. On July 11, 1890 the Kendallville Standard reported the following:

"The company will build a factory at once and push forward trade at all points. Besides their celebrated refrigerator and cold storage, they expect to manufacture a general line of butcher tools and supplies, and will stand second to no other firm in the United States in this branch of their business. The officers and directors of the corporation are gentlemen whose lives have spent in active business, and whose careers have been successful far above the average. They are just such men from whose exertions we may confidently expect a manufacturing equaling the Olivers and the Studebakers in volume and extent, and whose integrity will be as solid as the Bank of England. Kendallville may well congratulate itself upon the organization of the Company."

Before the patented McCray System, perishable foods were preserved by placing them in direct contact with the ice or in a box which was cooled by ice without circulation of cold air. The McCray patent involved arranging the ice-chamber so that the cold air (which, being heavier than warm air, always drops) will be in constant circulation throughout the refrigerator.

With capital of approximately $500 and a small building of 2,500 square feet, the first McCray products were produced in 1890. By 1948, the plant had expanded to 400,000 square feet extending over 9 acres.

McCray became one of the world's largest manufacturers of commercial refrigerators with branches in major cities throughout the United States.

From McClure's Magazine, 1906

From McClure's Magazine, 1899

61

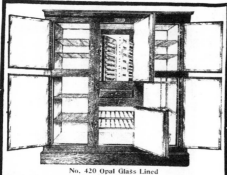

McCray Refrigerators

Porcelain Tile, Opal Glass or Wood Lined. All sizes, for Residences, Clubs, Hotels, Hospitals, Grocers, Markets, Florists, Etc.

Endorsed by physicians, hospitals and prominent people.

The McCray Patent System of Refrigeration

insures perfect circulation of pure cold air, **absolutely dry.** Salt or matches keep perfectly dry in a McCray Refrigerator, the most severe test possible.

Zinc Lined Refrigerators Cause Disease

That stale smell about a refrigerator is a danger signal. The zinc is corroding and the oxide poisoning the milk and food.

No. 420 Opal Glass Lined

McCray Refrigerators are lined throughout with Porcelain Tile. Opal Glass, or Odorless Wood **(no zinc is used).** They are Dry, Clean, and Hygienic, of superior construction, are **unequalled for economy of ice,** and can be iced from outside of house. **Every refrigerator is guaranteed.**

McCray Refrigerators are also Built to Order. Catalogues and Estimates Free. Catalogue No. 80 for residences ; No. 46 for hotels, restaurants, clubs, public institutions, etc ; No. 57 for meat markets; No. 64 for grocers; No. 70 for florists. **Book on " American Homes " sent free.**

McCray Refrigerator Co., 419 Mill St., Kendallville, Ind.

BRANCH OFFICES:

Chicago, 55 Wabash Ave.   St. Louis, 404 N. Third St.   Cincinnati, 326 Main St.   San Francisco, 122 Market St.
New York, 341 Broadway   Columbus, O., 66 N. High St.   Detroit, 305 Woodward Ave.   Minneapolis, 420 So. Third St.
Boston, 52 Commercial St. Philadelphia, 1217 Chestnut St. Pittsburg, 636 Smithfield St. Louisville, 421 W. Market St.
Cleveland, O., 64 Prospect St. Columbia, S.C., Hotel Jerome Bldg.   Washington, D.C., 620 F St., N. W.

Address main office unless you reside in one of the above cities.

From McClure's Magazine, 1905

Metal Stamping Company, Jackson, Michigan

White Frost Refrigerators

Safeguard the health of the family. More than 50% of disease can be traced to unwholesome wooden Refrigerators, which cannot be kept clean. The "White Frost" is all metal, not a splinter of wood about it, can't rust, warp, leak, decay. Enameled spotless white, inside and outside. No nasty corners for dirt or germs to lodge. Has revolving, Removable Shelves. May be washed out in a minute. It is always clean.

Keeps food pure and sweet, by natural refrigeration. Economical of ice. **Money back if not satisfied.** Send for free book telling about the most **perfect sanitary Refrigerator in the world.**

We will sell you one at trade discount, freight prepaid to your station, if your dealer does not handle them.

METAL STAMPING CO.
511 Mechanic St., Jackson, Mich.

Dear Bob, buy me a White Frost Refrigerator

Ladies Home Journal, June, 1909

According to the *1912 (Jackson) Directory* Metal Stamping Company was incorporated in 1904. A *Jackson Citizen Press* publication in 1912 stated "The chief product of this company is the White Frost and the Jack Frost refrigerators, sold in every country in the world. Nearly every army and navy post in this country is supplied with these refrigerators and they were also sent in large numbers to the Isthmus of Panama. The plant covers a space of seventeen acres and employs one hundred twenty-five men. Thirty salesmen cover every state in the union".

Apparently after 1912 the company either changed names or relocated as nothing could be found relating to it after that year.

Monroe Refrigerator Co., Lockland, Ohio

From McClure's Magazine, 1902

From McClure's Magazine, 1904

Ladies Home Journal, April, 1910

64

From McClure's Magazine, 1896

Northern Refrigerator Co., Grand Rapids, Michigan

The Northern Refrigerator Company was a division of Grand Rapids Refrigerator Co. It is presumed that Northern manufactured a lesser product than the parent company.

AS far ahead of all others as the *Electric Light excels the candle. Seven Walls* to preserve the Ice. Air-tight Locks. Dry Cold Air. Hardwood. Antique finish. Elegant designs. Sideboards or China Closets in combination, if desired.

Beyond question the Best Refrigerator Made.

Send for Catalogue. *We pay freight where we have no agent. Prices low.*

NORTHERN REFRIGERATOR CO.,
GRAND RAPIDS, MICHIGAN.

From Harpers Magazine, March, 1892

The Ranney Refrigerator Co., Greenville, Michigan

From starch to refrigerators! That indeed was the evolution leading to the founding of The Ranney Refrigerator Co. In 1890 the Greenville Starch Co. built a brick factory to make powdered starch from potatoes. A year later the company suspended operations because the price of potatoes had risen from $.70 to $1.40 per bushel, making it an unprofitable operation. On September 22, 1892 stockholders of the starch factory voted to sell the building to a refrigerator company.

The company was incorporated in October 1892 with capital stock of $50,000. The original officers were F.E. Ranney, President; C.T. Ranney, Vice President; F.A. Lamb, Secretary; and W.D. Johnson, Treasurer.

Greenville boasted of being the possessor the "the largest exclusive refrigerator plant in the United States". The company did indeed expand. They operated their own sawmills — cutting and delivering lumber from a radius of sixty miles. They also operated their own train of forty cars.

Photograph Courtesy Flat River Historical Society, Greenville, Michigan
The Ranney Refrigerator Company Plant

Ranney manufactured three distinct lines of refrigerators and over one hundred styles: the "Lapland" of solid oak, the "Monitor" made of solid ash, and the "Mascot" of other hardwoods. The list prices of their refrigerators ranged from $5.00 to $125.00.

During the 1930's the company recognized that the electric refrigerator was destined to replace ice box refrigerators and began to adapt to the change by selling cabinets to Westinghouse and Zenith on a contract manufacturing basis. By the 1960's the comapny was making a wide range of electric refrigerators and freezers.

In February 1970 the Ranney Refrigerator Company was acquired by Fedders Corporation. Prior to this sale the assets of the corporation were publicly owned by 412 stockholders. Management, since its beginning in 1892, had been in the hands of the Ranney family.

In 1978 Fedders Corporation withdrew from Greenville. A newly-formed corporation, Northland Refrigeration Company purchased the firm's Greenville operation.

Seeger Refrigerator Co., St. Paul, Minnesota

Ladies Home Journal, April, 1910

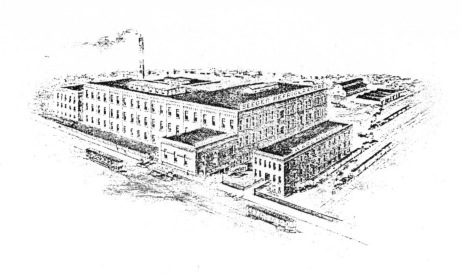

Our Guarantee

SEEGER Refrigerators are guaranteed to be exactly as represented. If found not as represented, we will refund purchase price and all freight charges. This guarantee is binding on us whether refrigerators are bought direct from us or one of our representatives.

SEEGER REFRIGERATOR CO.
ST. PAUL ··· U·S·A·

GENERAL OFFICES AND
MAIN FACTORIES

SAINT PAUL

BRANCHES
BOSTON, MASS., 82-84 Washington Street
DALLAS, TEXAS, 1307-1309 Elm Street
NEW YORK CITY, 101 Park Avenue

CATALOGUE TWENTY-FIVE

Cover from 1916 Seeger Refrigerator catalog

Our Drain System Our system of taking the melted water from the refrigerator is entirely new. The pipe and trap are located behind the front center rail, easy of access and can be cleaned easily and quickly—just lift out and clean as you would any dish. The drip pan is below the ice chamber and a pipe extends from it to the trap. The trap is covered by an aluminum cap which regulates the system so that no air can get into the refrigerator through the drip pipe. A second protection (as shown in figure twenty on Plate A) absolutely does away with every possible chance for air entering through the drain.

Elastic Enamel Lining Our elastic enamel is not mere white enamel but a specially constructed enamel applied with air and baked on to heavy galvanized steel by our own special process. It is snow white, as easily cleaned as a plate and positively will not chip, crack, peel, or be any shade or color other than white.

Porcelain Enamel Linings Our porcelain enamel, which is the very highest grade of hollow-ware, is applied to extra heavy sheet steel and fused on at a temperature of 2800 degrees. The fusing makes it practically a part of the steel. It is the most durable and best known material for the interior of refrigerators. There is no wear out to it.

Rear and End Icing Doors Keep the ice man outside and prevent the tracking of mud on your clean floors by having a *Seeger* Refrigerator with either an end icing door or a rear icing door. These doors are extra doors and stock sizes can be equipped with them. Complete details and measurements are given on pages 54 and 55.

Select your refrigerator and on receipt of your advice we will forward to you a complete drawing to be given your contractor, eliminating all chance for errors.

Solid Brass Hardware We have recently improved our door catches and handles, and *Seeger* Refrigerators are now equipped with a special combination, extra heavy self-acting lever with lock and key. Refrigerators Nos. 90-900, 95-950, are equipped with heavy lever latches. All hardware is solid brass.

Right or Left Hand Ice Chambers *Seeger* Refrigerators can be had with the ice chamber on either the right or left hand, as desired. All illustrations in the catalog show right hand ice chambers. All *Seeger* Refrigerators are shipped as shown unless otherwise ordered.

Germ Proof Shelves *Seeger* Refrigerators are equipped with specially constructed heavily tinned wire shelves, which rest on heavily tinned hooks and are easily lifted out for cleaning. Our elastic enamel lined refrigerators are equipped with tinned wire floor racks.

Finish *Seeger* Refrigerators are regularly finished Golden Oak and can be had also in different woods and different finishes in Stock Sizes or Specials at slight additional cost.

Water Coils Any *Seeger* Refrigerator can be had equipped with a block of tin water coil with a faucet and funnel, if for Bottle Water Service, plain pipe end if same is to connect to city water system. These coils are so made and so located in refrigerator that the ice consumption is not increased.

Features of Seeger Refrigerators — From the 1916 catalog.

Simmons Refrigerator Co., St. Paul, Minnesota

The general offices were located at No. 68 East Third Street, St. Paul, Minnesota. Their "Dry Air Refrigerators" were based on a patent issued January 18, 1881. Note the following diagram. When the front doors were opened the cold air was shut into the cold storage room by slides. When closed, the slides allowed the cold air to pass into the other compartments.

SECTIONAL CUT OF
Simmons' Patent Refrigerator.

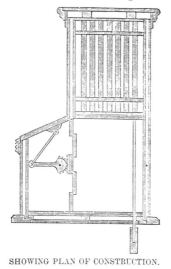

SHOWING PLAN OF CONSTRUCTION.

From Simmons Catalogue, 1888-9

From Simmons Catalogue, 1888-9

The company claimed the following features above all others:
- "They are a throughly dry air refrigerator, there being no condensation.
- They have a perfect cold air circulation.
- They can be run with seventy percent less ice than any other refrigerator made, of same capacity.
- Absolutely the only refrigerator in which can be stored butter, milk, fish, fruit, and vegetables, with anything usually kept in a refrigerator without contamination or transferring either taint or odor."

It is assumed that this company was purchased by or changed its name to Minnesota Refrigerator Co. around 1888-89. A catalog issued by Simmons for that period was modified accordingly.

Minnesota,
~~Simmons~~ Refrigerator Co.

SOLE MANUFACTURERS OF

SIMMONS'

WORLD-RENOWNED PATENT

REFRIGERATORS

—AND—

COLD STORAGE ROOMS.

GENERAL OFFICES,

No. 68 East Third Street,

ST. PAUL, MINN.

From Simmons Catalogue, 1888-9

E.H. Stafford Manufacturing Co., Chicago, Illinois

The E.H. Stafford Co. was established in 1900 and manufactured office furniture. Originally known as E.H Stafford & Bros. (E.H., Herbert A., and Edwin A.) the company was listed in the Chicago City directory at 17-23 Vanburen - 4th floor. The Stafford Company purchased the Caloric Company around 1919-1920, and the Stafford-Caloric Company was formed.

The Caloric Company, Janesville, Wisconsin, produced a fireless cookstove. The fireless cookers operated with heat — retaining soapstones which were preheated in a fireplace or in another stove. The preheated soapstone was then placed in the Caloric Cooker — supposedly saving fuel.

From 1920-28, Chicago city directories listed both the Stafford-Caloric Company and E.H. Stafford Company at the same address (367 W. Adams). The company apparently went out of business post 1928. No listings for the company or for Mr. E.H. Stafford appeared after 1933.

Jos. W. Wayne Manufacturer, Cincinnati, Ohio

The only information obtained regarding Joseph W. Wayne

71

were listings taken from the Cincinnati City Directories. Following are a few:

- 1866 Wayne, Joseph W., refrigerator and wash board manufacturer, 211 W. 5th
- 1871-72 Wayne, Joseph W., manufacturer of refrigerators, ice chests and beer coolers, 211 W. 5th
- 1883-84 Wayne, Joseph W., manufacturer of refrigerators, beer coolers and ice chests; also, Plimpton's patent roller skates, 124 Main
- 1896-97 Wayne, Joseph W., refrigerators, 326 Main
- 1898 Wayne, Joseph W., home 717 Gholson, Avondale

JOS. W. WAYNE, Manufacturer of the CELEBRATED

PATENT SELF-VENTILATING AMERICAN REFRIGERATORS

New and Improved BEER and ALE COOLERS, and all kinds of ICE CHESTS.

Ten First Premiums and Silver Medals Awarded at the Cincinnati Industrial Expositions.

| Twenty-five Years' experience has shown that they are undoubtedly | These are the only Refrigerators that are packed or filled with GROUND CORK between the walls, the best non-conducting material known for the purpose, as it does not absorb and retain moisture, and, being light and elastic, will not settle or pack down in the walls as charcoal or sawdust does. | the best, cheapest, and most reliable Refrigerators in the market, and |

Unexcelled for Simplicity, Efficiency, Economy and Durability.

Send for Illustrated Price-lists to JOS. W. WAYNE, Manuf'r, Liberal Terms to Dealers. 124 Main Street, Cincinnati, O.

From Century Magazine, May, 1888

JOSEPH W. WAYNE,

MANUFACTURER OF THE CELEBRATED PATENT SELF-VENTILATING

AMERICAN REFRIGERATORS,

IMPROVED BEER AND ALE COOLERS,

And Ice Chests of all kinds.

No. 211 West Fifth Street, between Elm and Plum Streets. - - - CINCINNATI, OHIO.

The importance and necessity of ventilation in Refrigerators, is now so well known, that it is unnecessary to enlarge upon it here.

FIVE YEARS EXPERIENCE has demonstrated the superiority of THE AMERICAN SELF-VENTILATING REFRIGERATOR, over all others, in the essentials of SIMPLICITY, EFFICIENCY, ECONOMY AND DURABILITY; while in style and finish, they are unsurpassed. They are therefore confidently recommended as

THE BEST REFRIGERATOR KNOWN!

From Cincinnati City Directories, 1869

To "Show Up" a Refrigerator

BLOW smoke into the bottom of the provision chamber of a refrigerator and you can SEE the air currents,—IF THERE ARE ANY.

You must look quick if it's a Bohn Syphon Refrigerator because the smoke will rush up to the top, and over into the ice chamber in a jiffy,—say in a third to half the time it takes in the "Next Best" Refrigerator.

Now, Mrs. Housekeeper, that certainly proves RAPID circulation of cold—real cold—air.

And hurry-up air circulation is the secret of refrigeration.

Because,—(now this will save you money and no end of trouble so read slowly,)—If the smoke you blow into the bottom of the

Bohn Syphon Refrigerator

quickly penetrates every corner of the provision chamber, and passes through the Syphons and into the ice-chamber in a third to half the time it takes in the pick of all other refrigerators,—

You can SEE with your own eyes that the air in a Bohn Syphon Refrigerator passes through the ice chamber two to three times as often—

And that it STAYS there a third to half as long. Now,—

Two to three times as often through the ice chamber means 10 to 20 degrees colder air in the Bohn Syphon Refrigerator.

It's easy to prove this—the thermometer tells the tale.

But, don't stop there—that's only half refrigeration.

The fact that it stays there only a third to half as long means that it has just that much less time to absorb moisture from the ice.

And moisture in a refrigerator encourages germ life almost enough to overbalance lowering the temperature.

And we all know that milk sours; and perishable foods spoil because of the multiplication of germ life.

And so unless your refrigerator both chills the air and keeps it dry, it's doing very little for you worth it's price.

Hang a wet cloth in your refrigerator and see if it will dry quicker than one just as wet hung in the warm room outside the refrigerator.

If it doesn't your refrigerator doesn't refrigerate. It will every time in a Bohn Syphon.

* * *

We can prove far better refrigeration. But you must see the Bohn Syphon Refrigerator to appreciate its beauty. No piece of finest furniture could be more beautifully finished.

And the Bohn Syphon is constructed of picked material and built for durability.

They cost $20.00,—and more for larger sizes. In every one large or small the Bohn Patent Syphons are the same—the construction and finish are identical.

SHE SAYS IT'S "THE BEST"

Ten Days Free Home Test

This will tell the tale. Our dealer in your town will deliver a Bohn Syphon Refrigerator of your selection for you to try. Keep it and use it 10 days. Just follow the printed directions on the ice-chamber door. If it fails to meet every test, tell the dealer to come and get it—you will not be out a penny.

If we have no dealer in your town we will ship you any refrigerator you select. If after 10 days actual use you want to return it, we will pay the return charges and hand back

every penny of your money.

We want to further prove that the Bohn Syphon Refrigerator will give better results at lower cost than any other.

May we send you our book, which illustrates our different refrigerators, gives a complete list of prices and explains more fully why the Bohn Syphon System is the very lowest cost refrigeration? Write us today—at once—it is really worth your while.

White Enamel Refrigerator Co., 1341 University Ave,, St. Paul, Minn.

From McClure's Magazine, 1906

Wilke Manufacturing Company, Anderson, Indiana

From McClure's Magazine, 1901

The Wisconsin Refrigerator Co., Eau Claire, Wisconsin

The Wisconsin Refrigerator Co., maker of the "Wisconsin Peerless" was started in 1888 with five small buildings on 2½ acres and grew rapidly to a 17 acre plant. Founding officers of the company were: W.J. Starr, President; G.W. Laurence, Vice President; F.H. Graham, Secretary; and F.L. Whetstone, Treasurer. By 1898 they were turning out 50 refrigerators a day, shipping them to all parts of the United States.

Eau Claire Cold Storage Corporation was an outgrowth of the Wisconsin Refrigerator factory. A disastrous fire in 1931 closed down the plant. The new company was formed and capitalized at $200,000.

A newly equipped factory had a capacity to produce 50,000 refrigerators a year. In the fall of 1932, they reported the best year in their history including an extraordinary sale of 1,000 refrigerators to the government.

The company became a casualty of the depression and closed down operations in 1936. It was reorganized the following year with new capital, but closed shortly thereafter. In 1943 the factory was purchased by White Machine.

Photograph Courtesy Chippewa Valley Museum,
Eau Claire, Wisconsin

A "Wisconsin Peerless" in a typical kitchen setting — early 1900's.

AN INVESTMENT?

An Investment?

An ice box as an investment? You bet! Over the long run appreciation of some antiques and collectibles have outpaced inflation. They also have out performed stocks and bonds over the last decade. Collectibles, ice boxes included, offer a unique opportunity for individuals to own a real "blue chip" investment.

Why have ice boxes appreciated in value? It is simply that they are a part of the Americana boom. Americans are searching for a piece of the past. They have discovered the richness of their heritage. Collecting artifacts produced by cabinetmakers, artists, and other craftsmen of the past has become a national pastime. William Stahl, head of the American Furniture Department at Sotheby Parke Bernet auction gallery, has stated "Each year since the bicentennial, has brought a dramatic increase in the number of new collectors in virtually all areas of Americana". Robert S. Salomon, Jr., a general partner in Salomon Brothers, a prestigious Wall Street securities trading firm, has stated "The market in collectibles now encompasses a spectrum of obscure items from beer cans to worthless — in an operating sense — securities. The breadth of interest suggests there is a limit to the downside risk in such investments."

Does this mean that ice boxes and other collectibles are immune from price declines? Not at all! In spite of the fact that many people think you can buy a collectible and it will just go up in value. Ice box prices are affected by a variety of factors: state of the economy, interest rates, regional preferences for furniture, fads, etc.

There are, however, some things you can do to make a wise investment decision (increasing the possibility of appreciation). Following are a few tips:

- Develop enough expertise to avoid having to depend entirely on the seller for information.

- Recognize there is disorderliness in the market. Valuing ice boxes, as with all antiques and collectibles, is an inexact science. The value of each item is determined subjectively.

- Recognize that the difference between bid and asked prices can be substantial — as much as 50%.

- Determine the regional preferences and price differences. Take time to shop when on vacations or business trips. For instance, prices on certain wood furniture in Grand Rapids, Michigan is going at about half the rates in New York and other major cities.

- Look for the uncommon. If you find two or three examples of the same ice box style or manufacturer in a shop, the chances are it will be, at best, an average investment. Buy one that's unique, one that sticks out in your mind, one that you like.

- Buy an ice box that needs refinishing or repairs — providing you are willing and able to do the refinishing or renovation yourself. If you are not, then buy one that has been reconditioned to your satisfaction. Refinishing is a time-consuming, frustrating job. It is also challenging and very rewarding.

An ice box can be an excellent investment. The tips offered above can help insure that acquiring, and perhaps refinishing one, will be fascinating and profitable.

Photograph Courtesy Julie J. Redd

This ice box was purchased for $90 in 1980. It was in terrible shape. Another $150 was spent on rebuilding, refinishing, and missing hardware. It is valued at over $600 today. Not a bad investment and a handsome piece of furniture to boot!

WHERE TO FIND THEM

Where To Find Them

There isn't a magic formula for finding and buying ice boxes. The key to successful collecting is exposure. Learn as much about them as you can. Go to antique shops, shows, flea markets, auctions, and museums. You are not required to buy and you can gain a great deal of knowledge. Examine pieces closely. Ask questions. Most importantly, make a note of prices, manufacturers names, and models that appeal to you. Finally, decide what kind you want and what you will have to pay for it. If you find an ice box you like, and it's priced right, buy it — it will have intrinsic worth to you and chances are it will increase in value.

Following are sources and tips for successful collecting:

Auctions

In America, auctions have been one of the best sources for antiques and collectibles for more than 200 years. Ice boxes are generally found at specialized antique auctions or country sales. The more sophisticated the auction — the less chance you have of finding a bargain. Fancy advertisements with photos of objects offered and catalogues cost money to produce. Unquestionably, these merchandising materials attract buyers. Which, in turn, puts upward pressure on prices.

Make no mistake about it, auctions are exciting. The competition of bidding is fun. So it's important to know how to bid and when to stop. Some important suggestions:

- Carefully inspect the box during the preview before the sale. Cardinal rule: don't buy unless you have thoroughly inspected the item.
- Don't be timid. Let the auctioneer clearly see you by raising your hand to bid.

- Try not to be noticeably enthusiastic in bidding regardless how much you want the item. Competitors may be influenced by your enthusiasm.
- Decide beforehand on what the box is worth to you and do not exceed that price when bidding. Stick to it regardless of the excitement.

Remember there are no "lay-away" plans at auctions. When it's over you pay your bill and remove the goods!

Auctions are fun and can be quite satisfying if you get the item wanted at your price.

Dealers/Antique Shops

Buying from a dealer has many advantages over alternate sources. It's convenient, he many times offers a wide selection of merchandise, and generally stands behind the purchase. Quite importantly, is the information they can provide. Most established dealers are knowledgeable and willing to share information.

Most dealers have their merchandise priced on the display floor. Although it is increasingly difficult to find dealers willing to "bargain" you should always attempt to purchase for less than the ticket price. Suggestion: Don't discuss lower prices of the merchandise unless you are seriously interested in the item. A favorite lead question for me to initiate the bargaining process is, "How much are you willing to take for this"? The worst that can happen is a refusal to budge from the list price. It has been quite common to obtain a minimum discount of 10%. Not bad for having asked a simple question.

Antique Shows

An antiques and collectibles show provides a marvelous opportunity to browse through displays of "old stuff". It is an efficient way to talk to several dealers and compare prices. Shows are also an excellent place to shop for a dealer.

Most big-city shows tend to have high standards. It is common for dealers to participate by invitation only. Many specialize in formal furniture and art. The chances of finding an ice box at one of these shows is slim, and if you do, the price will probably be exorbitant.

Smaller shows generally offer a much better opportunity for locating ice boxes since many offer less expensive "country items" and kitchen gadgets.

The dates and locations of shows appear in daily newspapers, trade publications, and collecting magazines such as the *Antique Trader Weekly* (check with your local library for a copy).

Flea Markets

Want some unique entertainment? Try a day at a flea market! Most flea markets are made up of individuals and dealers hawking a range of items from Elvis Presley records and used books to bicycles and ice boxes.

The best times to shop at a flea market is just as it is opening or right at closing time. The good collectibles at bargain prices go fast. Near closing times dealers commonly want to sell the merchandise that hasn't moved and are willing to bargain.

Prices at flea markets are frequently lower than those at an antique show or shop — providing you are willing to do some dickering.

So, visit a flea market and be prepared for an "experience"!

REFINISHING YOUR ICE BOX

Refinishing Your Ice Box

Finding the "right" ice box at a bargain price can truly be a satisfying experience. An even greater reward though, is the satisfaction of restoring one. It is not particularly difficult, but time, patience, and care are required. Carefully following each step of the restoration process will insure that the piece turns out as you want it. This chapter is devoted to the basic steps of ice box restoration.

Stripping

Most ice boxes were originally finished with shellac or paint. It is not uncommon, however, to find several coats that were applied during the useful life of the box. The following advertisement from the June 1898 issue of the *Ladies Home Journal,* and others like it, no doubt motivated housewives of that era to protect their ice boxes accordingly.

From Ladies Home Journal, June, 1898

Liquid paint removers are the best bet for stripping most ice boxes. Do one side at a time. That way you can position the box to insure having a flat surface which allows the remover to penetrate without running off.

If you are "scotch" by nature, be willing to forego your ways for this activity. Buy plenty of paint remover and apply it generously. Once applied, do not rebrush the remover. To do so weakens its effectiveness. It is very important to allow the remover to penetrate. Let the loosening (lifting) action take effect — normally five to fifteen minutes, depending on the finish and temperature. Normally the warmer it is — the quicker the "crinkling" effect.

When the loosening action has stopped, scrape the old finish from the surface with a putty knife. Work quickly, but do not scrape too hard. It is quite easy to scratch or gouge the newly exposed wood.

Frequently when stripping multi-finish coats, second or third applications are necessary. If multi-layers do exist, don't get over zealous in the scraping process. Let the remover do the work! One minor suggestion; put your "scrapings" in a used coffee can. The rigid edges will allow easy removal of the material from the putty knife. The plastic can cover helps insure that scrapings will be contained until the can can be disposed.

You will find there are always bits of paint that are not removed with a putty knife. Use steel wool for this purpose.

The last step is to cover the entire surface with a paint remover cleaner. This helps remove any remaining paint, varnish, and stain from the surface. Then rinse with clear water. Dry off the excess water and allow the wood to dry.

Stripping Steps

Generously apply remover to the surface. Do one side at a time.

There are three primary methods of stripping furniture: sanding, burning, and chemical removal. Regardless of which method used — it is only for the hearty! The objective of stripping is simply to remove the finish and expose the bare wood.

Each method has its disadvantages. Sanding is likely to remove more wood than is desired. Burning with a blow torch or special heater is inherently dangerous and likely to excessively darken the wood. With chemicals there are toxicity hazards from the remover itself and the paint being removed. Fire is also a danger.

The most effective method is chemical removal. There are many types and brands available but all fall into three basic categories: liquid, semi-paste, and paste. All are effective when used as directed. However, stripping is safer and easier if you use a non-flammable, non-grainraising remover that can be rinsed off with water.

Regardless of the type you choose, here are some common sense precautions to be observed in handling chemical paint and finish removers:

- Use adequate ventilation. Preferably work outdoors. If you cannot work outside, always work in a well ventilated area. Keep the windows open and use a fan if necessary.

- Wear rubber gloves. This will prevent absorption of solvents through the skin. Wearing cotton gloves inside the rubber ones can be helpful if your hand perspires excessively.

- Keep the remover off your skin — especially out of the eyes. Make sure there is running water available to flush away any accidental spills.

- Cover the work surface. Use newspapers, dropcloths, or plastic sheeting.

- Read the label. The use-directions and cautions will vary from product-to-product.

Let the remover set for 5-15 minutes — until the lifting effect has stopped.

Remove paint from the surface with a putty knife. Don't scratch or gouge the wood!

Use steel wool to remove paint you couldn't get with the putty knife. It is more gentle on the wood than sandpaper. And, it's important to protect the patina (mellowness) and texture of the old wood.

This is for the "big spenders". If after reading the stripping procedure you are discouraged to attempt the time-consuming messy job, there is an alternative for you. (I hope, however, you are even more motivated to tackle the job. Obviously you save money. More importantly is the self-satisfaction derived from the finished product. That is the great reward!) You can take your ice box to a professional stripper. Most utilize large tubs of potent finish remover in which the ice box is submerged. Using a professional stripper certainly eliminates this time-consuming messy chore. It will, however, cost considerably more than if you did it yourself. Prices will vary greatly depending on the size of the ice box and locality of the stripper.

Finishes

There are a variety of finishing products that bring out the natural color and grain in wood to insure it keeps its true beauty. I do not believe that staining ice boxes is appropriate and, therefore, have omitted any references in this chapter on restoration. More often than not, staining will make the ice box less attractive and less valuable. Therefore, these recommendations for finishing are designed to bring out the natural quality of the wood.

There are three catagories of finishes recommended: oil, penetrating wood sealer, and polyurethane. Each has a unique set of characteristics and are applied differently. Following is a brief description of each finish.

Oil Finish — It is an economical, permanent wood finish that accents the beauty of natural grain patterns. It is superior in wearing qualities and does not chip or peel. The finish and appearance actually improves with age.

To apply, simply dampen a lintless cloth with oil to penetrate the surface for approximately 30 minutes, then wipe the excess off completely. It's best to allow 6-12 hours between coats. Three to four coats are sufficient for most ice boxes.

Penetrating Wood Sealer — This finish provides a natural looking, hard, tough, and durable surface. Wood sealers are abrasion resistant. It apparently is due to the elasticity of tung oil or synthetic resins used as a base in sealers. The sealer is absorbed into and seals pores, saturates the surface, and becomes part of the wood.

Application is easy. Wipe the surface with a dry cloth before applying finish. Apply a thin coat using a lintless cloth (nylon fabric is good) to one section at a time. It is quick drying and, therefore, a must that excess sealer be removed immediately. Before applying the next and succeeding coats, allow at least 24 hours of drying time. Also, use very fine steel wool to smooth the surface. Be sure to wipe the surface with a dry cloth to remove fines and dust. Then proceed with the next coat.

Polyurethane — Clear polyurethane finishes make it possible to restore wood to its natural color. They are resistant to food, water, alcohol, grease and abrasion. This material does not penetrate into the wood. It forms a coating on the surface that is hard and extremely durable.

Application is a bit more difficult than with the other finishes. The surface should be cleaned of all foreign matter using a tack rag and a vacuum. Thin applications are recommended. Light steel wooling between coats with very fine steel wool should be done to assure the removal of small pieces of dust or foreign material which may have entered the finish when drying. Wipe clean with tack rag. The second coat should be applied in about 24 hours. No other finish should be used before or after applying the polyurethane.

Any one of the finishes recommended will reveal the natural wood color and grain. I personally prefer the polyurethane finish because of its durability and resistance to food and beverages. Before making your decision on which finish to use, read books on furniture restoration. Feel comfortable with your decision. Be sure to read the application directions of the product you buy very carefully. Each product varies in composition which can affect application techniques, drying times, etc. One final bit of advice on finishes. Buy a "name brand" even if it is more expensive. I've found the outcome is far more predictable. It's ridiculous to save a few pennies on the finish and risk the time and money you've spent on a project of this size.

The Hardware

Ice box hardware (hinges, latches, and manufacturer nameplate) varies greatly in size, shapes, and composition. Those made of brass or brass alloy are preferred because of its durability and appearance. Original brass hardware adds to the value of an ice box.

Much of the original hardware was nickel coated. I prefer to have it removed and strongly recommend using a professional.

Before Refinishing

After Refinishing

Restoration is not difficult. Follow the recommended steps and your project will be as successful as the one illustrated above.

One can be found in most cities. Many specialize in the restoration, cleaning, and polishing a wide variety of brass and copper items. Check the Yellow Pages in your area.

To clean brass hardware, follow these simply steps:

- Remove the hardware. Immerge and soak for 12-24 hours in a solution of equal parts household ammonia and water. Rub with very fine steel wool in warm soapy water. Rinse and dry.

- Apply a commercial brass cleaner. Some brands contain a low level of abrasives that aid in the cleaning process. A product like Brasso® will clean, polish, and protect brass.

- If the screws are corroded (you will notice when removing the hardware), replace them. Brass screws are easily found at a variety of stores.

What if an ice box is missing hardware? There are two options. One, visit antique shops that offer an assortment of "missing hardware". The chances are slim, however, that you will find a "match". I've been lucky in this regard and, therefore, should not totally discourage you.

The second option is to purchase replicas. There are companies that specialize in this. Two such companies are:

- Ritter & Hardware
  Gualala, CA 95445

- The Renovator's Supply
  Millers Falls, MA 01349

Restoration Summary

Restoration of ice boxes is not particularly difficult. It is tedious and time consuming. Make no mistake about that! However, by following a few simple steps you can avoid difficulties and have a totally successful restoration project. The basic steps mentioned in this chapter are not all inclusive. So obtain and study as much information as you can from other sources before beginning your project.

It is rewarding to restore any antique or collectible. You will get great satisfaction from your accomplishment.

Bibliography

"Belding-Hall Was Once Local Industrial Giant", The Belding Banner News, Belding, Michigan, August 29, 1957.

Bettner, Jill, "Personal Business", Business Week, March 24, 1980.

Business Week, "Liquidating Your Investment in Collectibles", June 11, 1979.

Carroll, William H., "When Ice Was Cut From Ponds for Farm Use", Sunday Sun Magazine, December 12, 1965.

Cummings, Richard O., "The American Ice Harvests", University of California Press, Berkley and Los Angeles, 1949.

Delco-Light Company, "Demonstrating Frigidaire, The Electric Refrigerator", Dayton, Ohio, 1923.

Demarest, Michael, "Going ... Going ... Gone!", Time, December 31, 1979.

Duis, Perry R. and Holt, Glen E., "Cold Times in the Ice Trade", Chicago, August, 1979.

"Elkins Refrigerators" (catalog), The Elkins Refrigerator and Fixture Co., Elkins, West Virginia, 1923.

Formby, Homer, "Formby's New Guide to Furniture and Home Care", Formby's Refinishing Products, Inc., 1977.

Gifford-Wood Company, "How to Harvest Ice", Hamilton Printing Co., Albany, New York, 1912.

Gifford-Wood Company, "Ice Tools Catalog No. 70", Hudson, New York, 1923.

Grand Rapids Refrigerator Co., "Leonard Cleanable Refrigerators", Grand Rapids, Michigan, 1914.

Hall, Henry, "The Ice Industry of the United States", Government Printing Office, Washington, 1888.

Harris, Maxene, "Professional Secrets Revealed for Ready-to-Finish Furniture", Missouri Ruralist, September 8, 1979.

Hiles, Theron L., "The Ice Crop", Orange Judd Company, New York, 1908.

"The Industrial Advantages of the State of Vermont", Commerce Pub. Co., Rochester, New York, 1890.

Jaeger, William J., "The Days of Ice Wagons and Ice Boxes", Sunday Sun Magazine, August 3, 1958.

Jobin, Judith, "How to Put a Price Tag on Your Attic Treasures", Woman's Day, May 13, 1980.

Kemp, Jim, "Arty Facts — How to Invest in Art & Antiques", Houston Home & Garden, February 1977.

Labine, Clem, "Hazards of Restoration Work", Chicago Sun-Times, 1979.

Littell, Alta L., "Industrial Romance Means Plain Hard Work", The Grand Rapids Herald, Grand Rapids, Michigan, July 25, 1926.

McCray Refrigerator Co., "Fifty Years of Progress (1890-1940)", Kendallville, Indiana, 1940.

Michael, H. Osborne, "When Baltimore's Ice Came from Maine", Baltimore Sunday Sun, March 15, 1959.

Minnesota (Simmons) Refrigerator Co., "Illustrated Catalogue for 1888-9", St. Paul, Minnesota, 1888.

Norwak, Nary, "Kitchen Antiques", Praeger Publishers, Inc., New York, New York, 1975.

O'Donnell, T.C., "Stocks? Or Collectibles?", Forbes, September 15, 1980.

"100 Years Ago", The Sun, January 8, 1967.

Rann, W.S., ed., "History of Chittenden County, Vermont", D. Mason & Co., Syracuse, New York, 1886.

Ranney Refrigerator Company, "Season of . . . 1897 . . . Illustrated Catalog", Greenville, Michigan, 1897.

Ranney Refrigerator Company, "1966 Annual Report", Greenville, Michigan, 1966.

Rick, Forest O., "Rhinelander Handbook of Refrigeration", The Rhinelander Refrigeration Co., Rhinelander, Wisconsin, 1926.

Seeger, "The Original Siphon Refrigerator", St. Paul, Minnesota, 1916.

Sherman, Harry C., "An Old-Time Iceman's Life", Sunday Sun Magazine, August 21, 1966.

Thames, Gena, "Furniture Restoration", Cooperative Extension Services of the Northeast States, NE-147.

Wearin, O.D., "Before the Colors Fade", Wallace-Homestead Book Co., Des Moines, Iowa, 1971.

White Enamel Refrigerator Co., "Why You Should Buy a Bohn Syphon Refrigerator", St. Paul, Minnesota, 1910.

"Wisconsin's Ice Provided Solid Base for Profitable Business in 19th Century", Wisconsin Then and Now, February 1975.

The World Book Encyclopedia, "Ice", I, Volume 10, 1974.

The World Book Encyclopedia, "Refrigeration", QR, Volume 16, 1974.

Younger, John, "Highlandtown's Only Iceman in the Early 1900's", The Sun Magazine, July 26, 1970.